PAPER MAKING
AS AN ARTISTIC CRAFT

by

JOHN MASON

illustrated by

RIGBY GRAHAM

FABER AND FABER
London · Boston

First published in 1959
by Faber and Faber Limited
3 Queen Square London WC1
First published in Faber Paperbacks 1979
Printed in Great Britain by
Whitstable Litho Ltd, Whitstable, Kent

British Library Cataloguing in Publication Data

Mason, John, b.1901
Paper making as an artistic craft.
1. Paper making and trade
I. Title
676'.22 TS1105

ISBN 0-571-11406-7

Acknowledgments

First to my wife for her forbearance when I used her kitchen, to my son Michael who helped forage for raw material, to J. W. Watson for loads of fibrous plants from Leicester Parks, to G. S. Percival for help with my early experiments, to Frank Tebbutt for long hours of hard skilful work, to J. Barcham Green for felts and much advice, to W. R. M. Watson, N. K. Kempt, K. Apold, Alex Ross and F. Goodger for never refusing my endless requests, to Amies for moulds, to Ron Bowers, D. M. Williamson, D. B. Beeby, G. E. J. Lowe for making all sorts of accessories, to the mills at Tuckenhay and Wookey Hole for materials and much advice, to Rigby Graham for his energetic help in practical artistic research, to the British United Shoe Machinery Company for the loan of a small beater, to Michael Chater and Francis Bolam for friendly counsel, to R. M. Harvey for making my larger beater and my sizing machine and vetting the text of this book, to Kenneth Holmes and E. E. Pullée for allowing my experiments at the Leicester College of Art, to S. J. Carter, J. A. Box, H. W. Fowler, R. E. Wood, R. D. Harding and J. C. H. Hurd for technical advice, to my colleague A. Pountney, the sculptor, for making the master for my portrait watermark, to St. Bride's Printing Library for allowing my exhibition there, to Dr. Desmond Flower for invitation to address the Double Crown Club, to David Bland for his sympathetic help in producing this book, to Sydney Walton and to Dr. Dard Hunter for the greatest encouragement, to Douglass Howell for real inspiration from across the Atlantic and to all the growing things that I transmuted.

Contents

Author's Note to the Paperback Edition

Paper Making as an Artistic Craft seems still to be the only simple practical textbook on this subject. Now, it is hoped, this new paperback edition will bring it within reach of all.

The various pieces of equipment and the techniques described are my own personal answers to the problems that arise. You may well devise others. Be adventurous and experiment yourself. Make that your charter. Many discoveries lie just around the corner in this most unexploited craft.

Pulp plants from your garden and from the river banks, the field and the hedgerows and even pulp your own shirts and frocks.

Paper with all sorts of natural colours and unusual textures will result—good to look at and good to feel. After a little practice you will make sheets suitable to use for drawing and painting and for hand printing too.

Express yourself through this work of your hands. Books which I have printed on paper which I have made now reside in many famous libraries around the world. One day, perhaps, they will become my small memorial.

JOHN MASON
Leicester 1979

Foreword

Several years ago The Institute of Paper Chemistry received a letter from John Mason of Leicester, England, marked for the attention of the Curator of the Dard Hunter Paper Museum, housed in the administration building of the Institute. The Paper Museum embraces an assemblage of 'paperiana' gathered in all parts of the world, showing the evolution of this craft from its invention by the Chinese eunuch, Ts'ai Lun, about A.D. 105, to the most modern methods of making paper by hand. It may seem an anachronism that the most up-to-the-minute papermaking school and laboratories in the United States should deem it worth while to sponsor such an archaic museum. But, the officials of the Institute were convinced that a knowledge of the history of paper would be decidedly beneficial to everyone concerned with present-day paper making by the most modern mechanical methods. It must be remembered that for almost 1,700 years there was no paper in the world other than that made by hand. The paper-machine is a comparatively recent invention, perfected only about one hundred and fifty years ago.

But, to return to John Mason: we, at the Institute, were all interested in his experiments in making paper by hand in his

'Twelve by Eight' mill and in the College of Art, Leicester. John Mason was making sheets of paper upon a mould that measured 12 by 8 inches in an entirely unorthodox manner, even from the standpoint of the accepted methods of producing commercial handmade papers. His wholehearted desire was to make paper of pronounced aesthetic quality, with no thought of quantity production, or in imitation of any paper that had ever been made.

To arrive at his quest for papers of unusual appearance, formation, colour and texture, he gathers his materials in the English country-side and cooks and macerates the vegetable fibres in his own miniature laboratory. He makes use of nettles, cow-parsley, rushes, grasses, and the stalks and leaves of such flowering plants as gladiolus and iris. We were at once reminded of the work of Dr. Jacob Christian Schaeffer (1718–90), a clergyman of Regensburg, Germany, who as early as the middle eighteenth century began his experiments in making paper from many forms of vegetation found growing near his home. Here he set up a small hand-operated stamping-mill for beating the material, and also a vat for forming the small sheets of paper. Unlike John Mason, whose principal desire is for artistic feeling in his papers, Dr. Schaeffer's chief concern was to find inexpensive materials that could be substituted for the scarce and costly linen and cotton then in use. Dr. Schaeffer compiled six pamphlets, from 1765 to 1771, in which he wrote of his many experiments in making paper from hemp, straw, cabbage leaves, asbestos, cattail and burdock stalks, thistles, mallow, St. John's wort, Indian corn husks, genista, pine-cones, potatoes, reeds, bean, horse-chestnut, walnut, tulip, and linden leaves, in all almost eighty different substances. This eighteenth-century treatise, with the many small original specimens of paper, is the rarest and most desired item in the entire bibliography of paper making.

During the past ten years there has been an increasing interest in the age-old craft of paper making from a purely artistic standpoint and a number of amateur paper makers have sent specimens of their work to the Paper Museum. This is indeed a most welcome and encouraging trend, especially in these prosaic times when every industry is so thoroughly dominated by trade unions and mass-production methods.

Knowing of John Mason's background which has recently been revealed to us, it is not difficult to trace how he became interested in the art of fine printing, which in turn naturally led to the more obscure craft of making paper by hand. Notable and artistic talents can usually be traced to distinguished heredity and environment, and of these John Mason had an abundant share. His father, J. H. Mason (1875–1951), was one of the enthusiastic workers in the English private press movement which had such a marked influence from the latter nineteenth century onward. The printers and book collectors of Europe and America were inspired by the books produced by the Vale, Doves, and Cranach Presses, in the making of which J. H. Mason contributed his artistic talents and skills. He also studied and worked with Sir Emery Walker, probably the most adept of all European typographers, who, more than any other individual was responsible for the books printed by William Morris and the Kelmscott Press.

We need more workers like J. H. Mason and his son, for only through painstaking craftsmanship can the world hope to survive beyond the monotonous mechanical existence that predominates at the present day. It is hoped that John

Foreword

Mason's book will prompt other artists to take up the absorbing diversion of making their own paper through using the simple and direct methods of the old paper makers.

The Dard Hunter Paper Museum,
The Institute of Paper Chemistry,
Appleton, Wisconsin, U.S.A.
20*th October* 1958

DARD HUNTER,
Honorary Curator

Reed-mace at Aylestone, Leicester

Preface

Paper making has become a highly mechanized industry serving the requirements of modern life. Whole forests are pulped and papers of all kinds stream from monster machines. A paper maker is now a highly skilled technologist, chemist and engineer. He manipulates valves and takes intricate readings but, except in hand-made mills, he has little personal contact with his product.

There are now only three small mills making hand-made paper left in England. This is alarming for soon there may be none. Hand-made paper is a very special material, delightful to see, delightful to use and wonderfully durable, too, as a copy of the Gutenberg Bible will show. At all costs this hand craft must be preserved. To make full-size sheets of a standard quality considerable skill and strength are required and some heavy equipment. On the other hand, I have found that small sheets are easy to make and simple apparatus quite easy to devise. For this there is no need to be a chemist or to spend much money. Plenty of raw material is available in the garden, under the hedges and in the rag bag. To share with you some exciting experiences in such paper making is the reason for this book.

CHAPTER I

A Synopsis

Natural animal and vegetable fibres have been spun into threads and woven into cloth since very early times. The disposing in layers of short vegetable fibres to form sheets of paper, however, was a much later discovery usually credited to Ts'ai Lun in China about the year A.D. 105. Recent finds in China indicate that paper may have been in use even as early as the year 140 B.C. Most natural fibres are too long and too thick to use as they are for paper making and they have first to be shortened and split lengthwise into fine fibrils. When a piece of bamboo is hammered upon a hard surface it splits up into many fine strands. A similar method of pounding, known as beating, is used to reduce fibres for paper making.

Only vegetable fibres can be split in this way and animal hair is of no use at all to the papermaker. Pure vegetable fibre in a usable raw state is found nowhere but in the seed pods of the cotton plant. Nevertheless, nearly all plants have a more or less fibrous cellular structure and in many it is sufficiently tough to be made into paper. In the more fleshy varieties the proportion of fibre is too small to be worth the trouble of separation. Different plants give very different characteristics to paper made from them and this affords a whole field for exciting experiments.

A preliminary boiling in a solution of caustic soda separates the fibres from other unwanted material which is then washed

away. The wet mass is beaten to pulp by pounding with heavy hammers or is repeatedly passed beneath a heavy revolving roller. When sufficiently reduced, the pulp is transferred to a vat and then thinned out with water. If the vat can be heated to keep the pulp warm some gelatine size can be added too.

The sheets of paper may now be formed by means of a *mould* and *deckle*. The mould is a wooden frame over which has been stretched a piece of woven copper wire to form a sieve. The deckle is a second frame, without wire, which fits over the mould and serves as a loose rim to confine the pulp. Mould and deckle are dipped into the vat and, with a scooping movement, a small quantity of pulp is picked up. The water is allowed to drain away, the deckle is removed and a sheet of paper lies glistening wet upon the mould. Carefully the mould is turned over and pressed, face down, upon a wetted blanket. On lifting, the paper stays behind and is covered by a second blanket. The operation is repeated, paper and blanket, paper and blanket, until a pile is formed about ten inches high. A wooden board is placed at top and bottom and the whole is screwed tightly into a press to remove the water. Removed from the press, carefully separated, the tender sheets are laid out to dry. When size has not been added to the pulp in the vat it is applied later to the sheets in a separate bath and the process is known as 'tub sizing'. After a final pressing the sheets are allowed time to mature and are then ready for use.

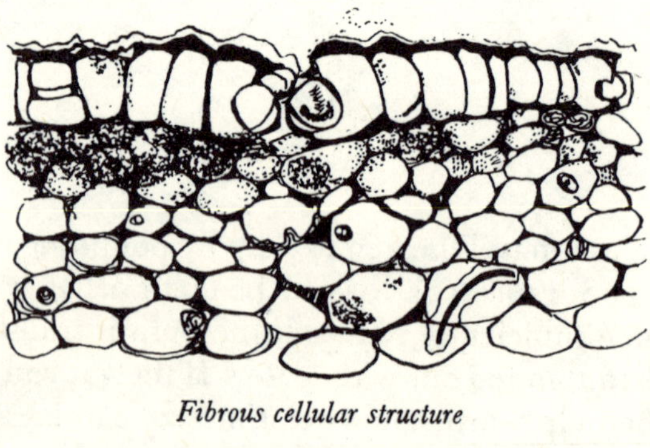

Fibrous cellular structure

CHAPTER II

Preparation for Beating

I first began my paper making in the kitchen, but soon found great disadvantages in this. Saucepans boiled over and caustic soda spoiled the enamel of the stove. Often I was in the way and made too much mess. In some households it might work, especially if the whole family shared in the enterprise. An outhouse, however, is much to be preferred and should be requisitioned if at all possible. When this has water laid on, gas and electricity, too, then conditions are ideal. For a start you will need two or three saucepans of various sizes and a pint and a quart measure. Stainless steel is best, but not essential. It is important to remember that rust is the great enemy and nothing that can rust should be brought into contact with the raw material, pulp or paper. A set of sieves from $\frac{1}{8}$ inch for the coarsest down to those of fine nylon will be required, but start with three and add as you gain experience of your needs. Stone or earthenware jars, and bins as used for bread, are best for the storage of raw materials and glass jars for chemicals, all with lids. Two or three buckets, preferably of the plastic type, will be needed. Buy one pound of caustic soda and half a pound of bleach powder and transfer them separately to stone or glass containers. Place a cupful

Common nettle

Flax, one of the earliest fibres used by man

of the bleach powder into another jar, add a quart of cold water and stir well at intervals with a stick. This will ensure that some bleach liquid is ready when you need it.

You may now prepare some raw material ready for beating. Towards the end of the summer or in the autumn you would do well to start with some wild plants such as nettle, cow-parsley, leaves of rushes and coarse grasses, and several kinds may be mixed together, if you wish. Leaves of garden plants such as montbretia, gladiolus and iris, also make excellent paper. Surprisingly large quantities are needed to make even a small amount of pulp and so a good large sack should be filled and pressed down tightly. Later you will find that the fibres are separated more easily if the plants are placed for a time in a ditch or in a tank of water so that they partly rot, or they may be left in a wet mass on a concrete floor. This is similar to *retting* of flax in Ireland in preparation for the making of linen. At first, however, this process may be omitted. The raw material is now cut up in $\frac{1}{2}$-inch lengths with a large pair of scissors or shears or by means of a card-cutter,

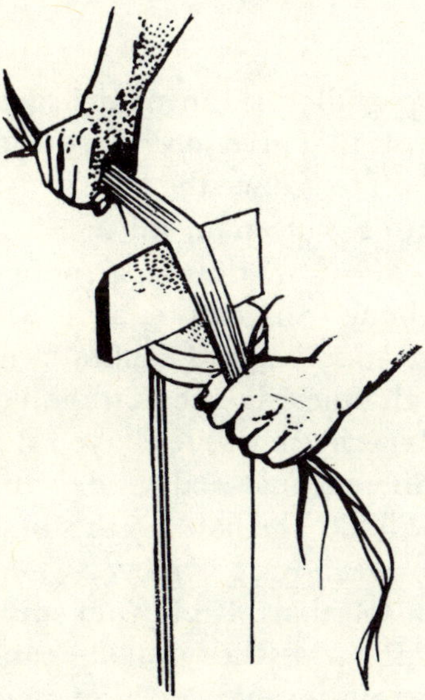

Breaking up leaves of iris

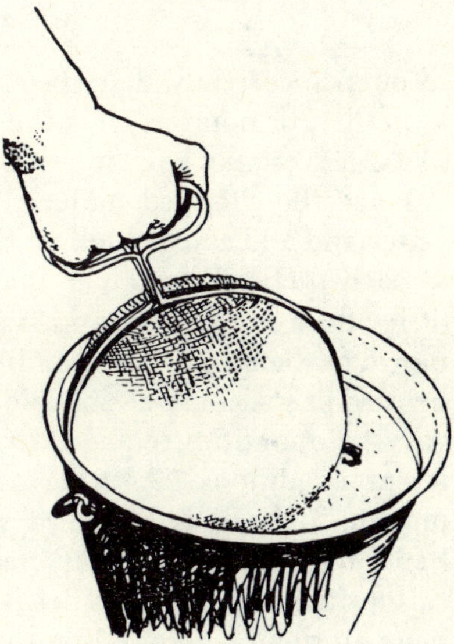

Straining out lumps

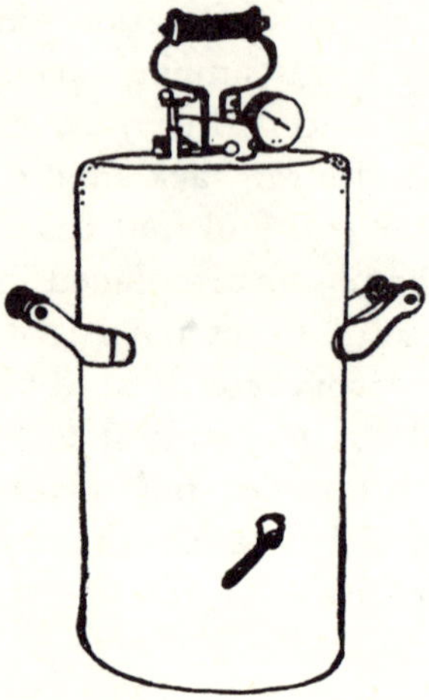

so often incorrectly described as a guillotine. In plants such as nettles, or potatoes, it is the stems that give most fibre and so the leaves may be first stripped off to reduce the bulk.

Place the chopped material into a saucepan, fill up with water and add caustic soda at the rate of two dessert-spoonfuls to each quart. Remember that though the caustic soda has little effect on the fibre and is used to soften and facilitate removal of the fleshy parts, yet like all other chemicals it should be used as sparingly as possible. Experience will teach you the varying quantities required by different materials under different conditions. 'As little as will do the job' is the best maxim. It is advisable always to wear rubber gloves when handling caustic and other bleach solutions. Bring your mixture to the boil and then let it simmer, as with cooking cabbage, until all is soft and pulpy. Strain through a coarse sieve and then proceed to wash away the non-fibrous debris. Water

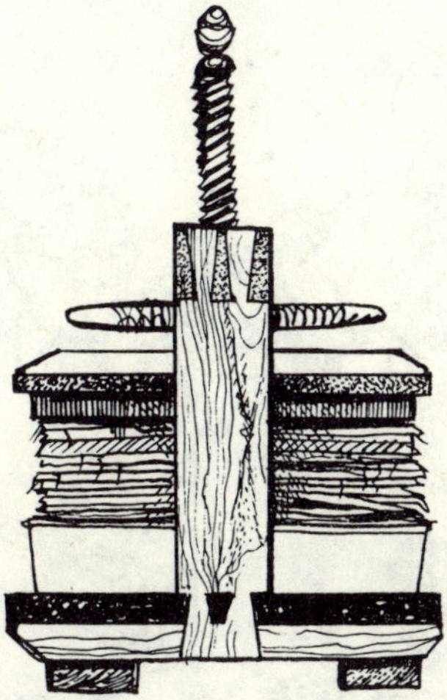

from the tap should play directly down as you twist and turn the mass with your hands. Transfer part into a smaller sieve and finish the cleaning process by plunging the sieve into a bath or sink full of tepid water. The suction caused by this action clears away the smaller pieces. Place the fibre into a stone jar, and then fill up with water to which a cupful of bleach liquor has been added. If you like your paper to have the same natural tint as the raw material, then there is no need to use any bleach at all. Unbleached paper is always the strongest and so the less the bleaching, the better. Leave the pulp standing overnight, even longer if possible, and then thoroughly wash again in a sieve to remove the bleach liquor. Use only a fine-mesh sieve from now on or you will wash away some of the precious paper-making fibre.

Where no suitable garden or field plants are readily avail-

able or if you are starting work in the period from winter to early summer, before plants are sufficiently mature to harvest, then make your first paper from rags. You must sort these carefully, selecting only those made from cotton. Cotton rags are the chief constituent of professionally hand-made papers, though linen rags, also, are used to some extent. Linen is more difficult to pulp, however, and it should be left until considerable experience has been obtained in handling materials more easily worked. Reject all rags composed in whole or part of wool or nylon. White and near white cotton rags will be kept apart for your white paper and will need very little bleaching. The coloureds may be sorted separately and, if you have a quantity of any one colour, paper of the same colour may be made without any bleaching. The remaining rags, of miscellaneous colours, are mixed for making up into paper of a neutral shade. Remove buttons, fasteners and similar things, and cut away all stitched seams. Then cut up into pieces roughly 1 inch square. Boil the rags for at least five hours in a solution of caustic soda in order to remove any dirt and to loosen the weave. The liquor is sufficiently strong when it imparts a slightly greasy feel to the fingers. If an old pressure cooker is available this is better for the boiling than a saucepan. Drain, wash thoroughly and tip in a pile upon a stone floor or in an enamel bath and leave to *ret* for as long as possible before proceeding with the *beating*.

CHAPTER III

Beating to Pulp

The reduction and fibrillation of the raw material for paper making is achieved by some form of pounding. It is best to start with a very simple method and then to enlist whatever mechanical aid you can devise. But first the

Japanese beating mallets

material should have a preliminary bruising and shortening by passing it through a heavy duty mincing machine. I used

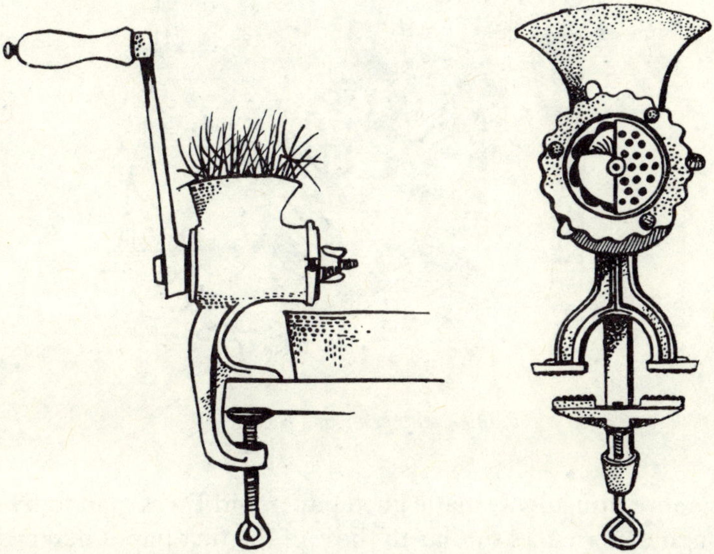

an ordinary pestle and mortar for making my first pulp. It was very hard work, and slow, and I was fortunate to have selected rather unripe cow-parsley, or it might have been

Pulping with pestle and mortar

even more difficult. It made good paper and I was glad to live through experiences similar to those of the first paper makers, but I soon looked around for a quicker method. Eventually I found the old handle operated mechanical mortar shown in the drawing. My output was more than doubled. Then I tried pounding with a mallet inside a stone trough and this,

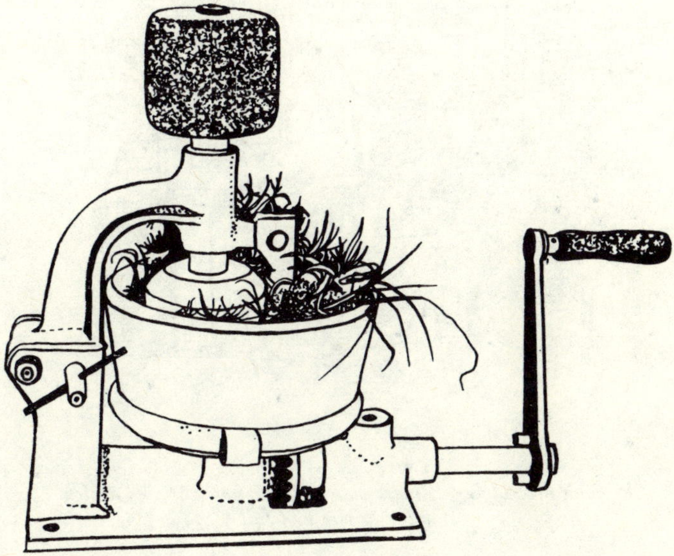

A mechanical mortar which I discovered one day in a junk store. Other suitable devices may be found to lighten the labour of pulping

though the work was hard, proved very effective. In Japan, balls of wet fibre or kozo are still beaten down to fibre with oak batons in this way. The old mills throughout Europe used a series of water-driven mallets or stampers for making their pulp. My illustration on the next page is of stampers which are still in action in the ancient mill of Richard de Bas at Ambert in France. Small mills in India still use a similar

Pegs on a revolving cylinder cause the hammers of this ancient stamper to rise and fall

machine called a *dhenki*, the stampers of which men operate with their feet.

A friend of mine lent me a machine on the same principle which he had made to test the wear resistance of army socks. There must be many such contrivances lying idle which could be adapted for making pulp. I worked day and night with this machine for the short time of the loan. After that I felt that I just could not go back to the pestle and mortar.

During the summer of 1954 I had a real stroke of luck through giving a luncheon talk on paper making to Rotarians at Loughborough. I expressed the hope that one day I could afford a miniature modern beater. 'Then I can speed up and

Dhenki in use in Southern India

African women preparing pulp

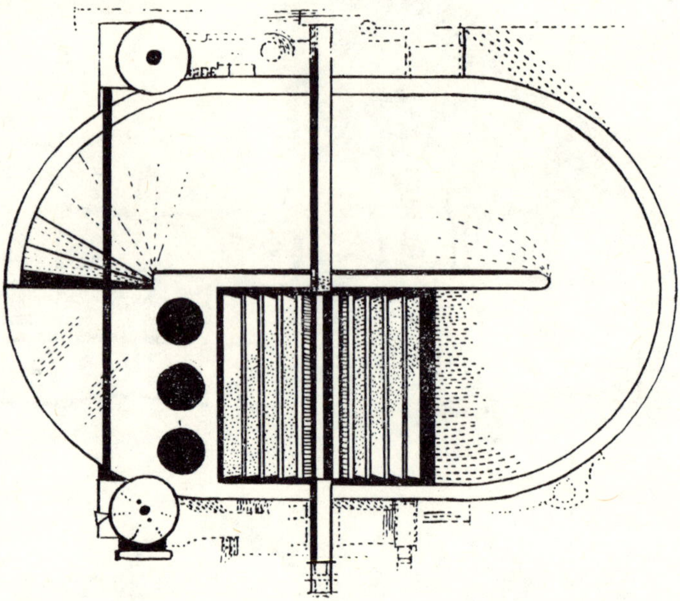

Top illustration is looking down on the beater with cover removed. Below is the usual appearance of beater with motor drive and roll cover replaced

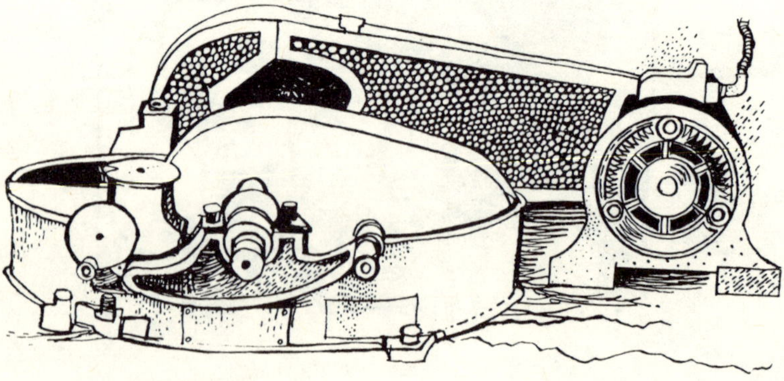

End view of beater roll showing movement of pulp during fibrillation

perhaps make paper from nylon', I told them. My audience was interested and promised help. Within a few weeks I had my model beater, discovered disused in the research department of a large factory. This beater is a miniature model of the Bertram Hollander found in most paper-mills. It is made of bronze and consists of a round-ended trough partly partitioned lengthwise to form an oval track and round this travels the pulp. This partition is called the *midfeather*. A heavy beater roll, bearing thick projecting bars, is mounted on a centre shaft on one side of the midfeather. As the roll turns it circulates the pulp around the trough. Below the roll is a bedplate, also carrying projecting bars. The pulp passes between roll

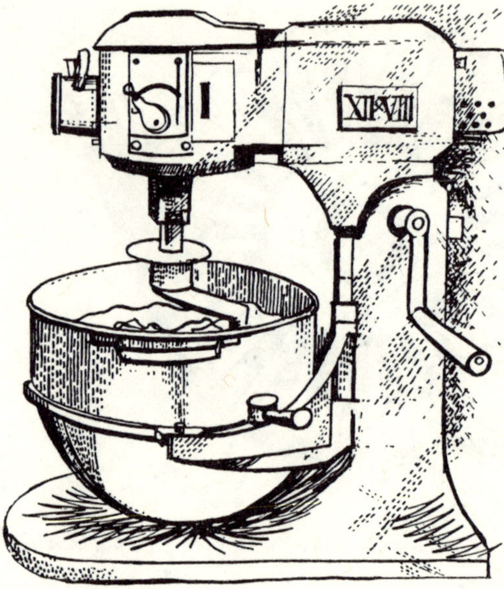

A food mixer is invaluable to the small paper maker. This one is power driven and I bought it cheaply from a village baker

and bedplate and is by this means cut and crushed according to the amount of clearance allowed. The quality of the paper is largely determined in the beater and the length of beating time and the amount of pressure from the roll are all-important factors. This little beater, however, makes only a few pounds of pulp at a time and so my rate of production was still quite slow.

Later I obtained an old Crypto food-mixing machine. This

gives the minced material a pre-beating mix which has greatly reduced the beating time. A professional paper maker recently built a second beater for me of another kind. This is known as the Harvey Beater and is of the Umpherston type. Though

My friend Roland Harvey, paper maker with long experience, built this beater specially for me. It has stepped up my production but I still use the smaller beater for small quantities

The boiled fibre has been well stirred in the mixing machine and then is poured into the beater. The roll cover is replaced and the two hatch doors are closed before the beater is started up

similar in action to the Bertram, it circulates downwards into a tank of large capacity and so makes much more pulp at each filling. Several times during the beating a pinch of pulp is mixed into a glass beaker of water and held to the light. It is easy then to see when all lumps and coarseness have been sufficiently reduced.

The addition of paper-makers' starch, a teaspoonful to a gallon of pulp, improves the formation and wet strength of the sheets.

It is possible to make some exciting experimental papers from pulp that is only partly beaten and still contains small unbroken pieces of grass or woven cloth.

Women in Japan beating pulp for making paper by hand. Throughout the night the hills resound with noise from this laborious work

Charming papers may be made on cloth moulds but the sheets are thin and the process is slow

CHAPTER IV

The Mould

The mould is the most important piece of equipment of the paper maker for with this he actually makes the sheets. It consists of a rectangular wire strainer with an outer wooden frame. Moulds and paper of other shapes seem never to have been required though it would be interesting to try a variation for an envelope or table mat.

For making the first papers it is thought that frames covered with woven cloth were floated on water and pulp poured over them. Upon lifting, the water drained through the cloth and the mould was placed in the sun until the sheet of paper dried and peeled off. A cloth-covered mould may be used also with the more modern dipping method when making experimental

Laid mould

papers. Sheets made in this way without a deckle and left to dry upon the mould are usually very thin at the edges and thicker in the middle. Soon cloth was discarded and a strainer made by stretching very fine strands of bamboo fibre or tough grass across the frame as close together as 20 to 30 strands to the inch. These *laid* lines were securely tied by cross lines of tiny chain stitches of silk or horsehair made at 1-inch intervals. With one *laid* mould, sheets could now be transferred immediately, while still wet, to a quickly growing pile. Except that we use bronze or brass wire, the method is similar to that employed today.

I decided on 12 × 8 inches as the size for my first paper, and Amies of Kent made the laid mould and deckle complete with my name as watermark. Watermark designs are formed in fine wire and sewn with still finer wire to the upper side of the mould. As they are in relief, less pulp lies upon them than over the rest of the mould. The resulting thinness gives the familiar transparency to the watermark when the paper is held to the light. Watermarks should be in fine line and not too prominent, otherwise they may cause imperfections of printing.

The following are details of my first mould. The frame is of mahogany in $1\frac{1}{2} \times \frac{1}{2}$-inch strips and measures $9\frac{1}{2} \times 13\frac{1}{2}$ inches, outside measurements. The corners are mitred, mortised and screwed. The frame is made firm and rigid by $\frac{3}{4} \times \frac{1}{4}$-inch wooden supports fitted at 1-inch intervals and level with the lower edge of the frame. These supports taper away to a thin bottom edge and are fixed to the frame by short pegs.

Deckle

Each support bar is pierced with a series of fine holes $\frac{1}{4}$ inch apart. The laid wires, spaced at 22 to the inch, are sewn with very fine wire to these supports and form lines of chain

The simplest form of mould is a plain wooden frame over which oddments of woven cloth are stretched. No deckle is used and the sheet is allowed to dry upon the mould

stitches 1 inch apart. The laid lines and chain stitches give the familiar watermarking seen in all laid paper. Half-inch strip copper is nailed round to cover and bind down the ends of the wires. The deckle frame is made from $1\frac{1}{2} \times 1$-inch mahogany, shaped as shown in the drawing and with the short ends protected with sheet brass. The deckle, which confines the pulp and determines the exact size of the sheet, fits loosely for quick removal.

All paper was made upon laid moulds until the middle of the eighteenth century when John Baskerville, the printer, demanded perfectly smooth sheets without laid lines. This was

achieved by the use of fine woven wire cloth in place of the laid wire. So smooth wove paper was born.

I have not attempted to make a laid mould myself, but I have made several good ones of the wove type. The mahogany frame with supports is as previously described. The professional wove mould has a series of strong wires stitched at

If you have an oriental kind of patience a simple wooden frame, covered with cloth, forms an excellent mould. Couching is not then possible, however, and you must wait for each sheet to dry before proceeding

$\frac{1}{4}$-inch intervals across the supports. A sheet of fine woven wire is then laid over, pulled taut and tacked at the edges which are then covered with copper strip. I have found that if coarse woven wire with $\frac{1}{8}$-inch mesh is nailed as a support below the fine top wire, no sewing at all is needed. For my top wire I obtained a length of discarded wove machine wire from a paper-mill, but any woven brass wire of about 30 mesh will serve as well. To ensure that the water gets away freely, drainage holes should be pierced in the wire along all four edges at about $\frac{3}{8}$-inch intervals. The shaping of a deckle in mahogany was too difficult for me and so I had several cast in aluminium. Even a crudely made wove mould and deckle, just held together with brass or copper nails may make good paper, but only the best materials and sound construction will ensure that they give long service.

There are many plants in the countryside which are yet untried for paper making. What better souvenir of a holiday than a sheet or two of paper which includes a few wild flowers or grasses

Good raw material for paper making may be found at the edges of canals, streams, rivers and ponds. Try collecting at different times of the year and compare results

CHAPTER V

The Vat

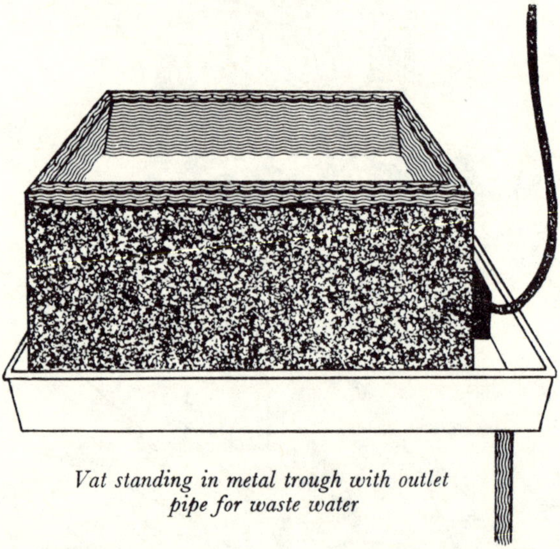

Vat standing in metal trough with outlet pipe for waste water

In early times the container for the pulp was a sawn-off half of a very large wine or beer cask. Nowadays a rectangular tank is used but it is still known as the *vat.* A small kitchen sink will serve for early experiments.

My first vat was a square box, with sloping back, made from pine. This leaked a little and also I found that I had not

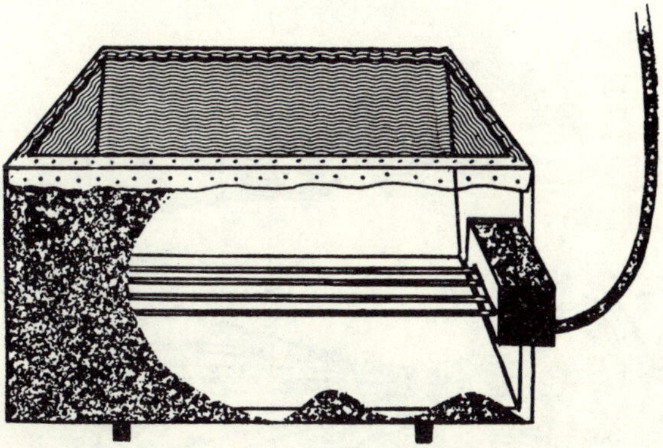

I made a stirring device of revolving spindles but it was not a success. Now I stir with a stick. This vat is 2 feet wide, 18 inches from back to front and 1 foot deep—just a comfortable size for working with a 12 by 8 mould

left enough room for easy movements with the mould. The improved vat shown in the drawing is of teak and the size is just right for making paper 12 × 8 inches. Once this had been thoroughly soaked there were no leaks. Later these wooden vats were lined inside with sheet copper and this was well worth the extra cost. Plain rigid copper tanks are now obtainable and are excellent for paper making if not quite so convenient as the deeper wooden ones with sloping backs.

Though paper can be made from cold pulp it is much better to work with it warm. The water then drains from the mould more quickly and the vatman is less likely to suffer from chilblains. A jug of boiling water added occasionally will keep things right. An electric heating plate fitted into the base is a refinement which rivals the steam pipes used in the professional vats. The professional also has a stirring mechanism known as the *hog* which keeps the fibres moving in suspension and pre-

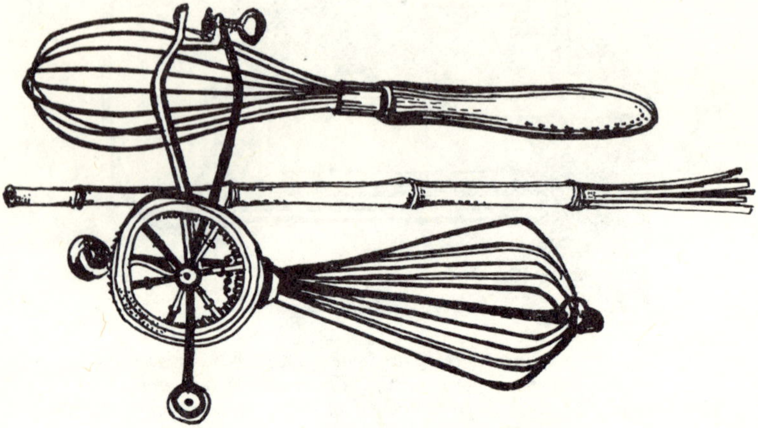

vents their settling to the bottom. After many experiments, however, I am convinced that the best method for us is the more simple device of stirring occasionally with a wooden stick or cane whisk.

Slopping water and leaking pipes create the right atmosphere for the vat-house of a paper-mill and I do hope that the works efficiency and motion study people will never hear about it. Nevertheless, we should keep down water spilling ourselves for mopping up is a non-productive operation. Dipping and lifting the mould inevitably causes a few splashes and so the vat should be stood on two teak battens, inside a shallow trough made from sheet zinc. The water which collects in the trough can be siphoned off with a rubber tube or, better still, a permanent waste pipe may be fitted leading straight to a bucket.

A disaster once befell me. House-proud, a friend in my absence cleaned my copper-lined vat with wire wool. The embedded iron particles caused small brown spots, known as 'foxing', to appear in all my paper and I had to burn it. Acids were of no avail and extraction by a powerful electro-magnet was the only cure. Vatmen beware. Paper making is a mystery, a magic. Once we stole it from the wasps and they are still taking vengeance.

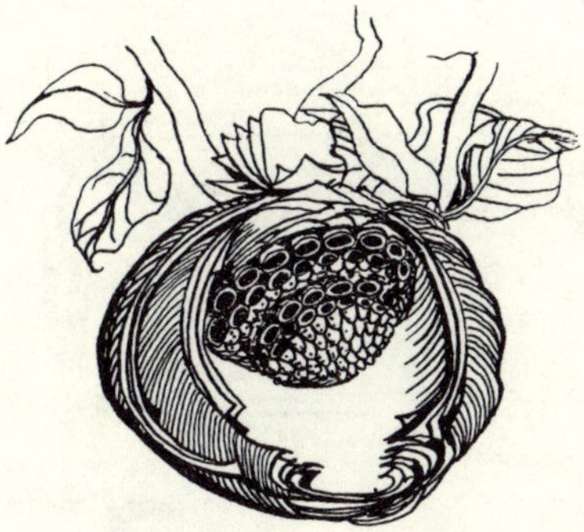

CHAPTER VI

Pressing Equipment

In order to consolidate the fibres and to facilitate handling and drying, it is necessary to squeeze out as much as possible of the free water from a 'post', the professional term for a newly made pile of sheets.

Pressing Equipment

First a wooden board, preferably of teak, is placed at top and bottom of the pile. A good deal of water may then be forced out by pressure of the hands or by piling up heavy weights of stone or metal. In parts of Japan a simple but ingenious lever device is still used very effectively, and I understand that in village mills in India workmen dance up and down on top of the pile as is done when treading grapes. A screw press, however, acts as a mechanical equivalent of piled-up weights and some type of screw press should be obtained as soon as possible. A bookbinder's wooden lying press, actuated by a screw at each end, is quite good and the water may be caught in a tray underneath. Old oak linen presses and

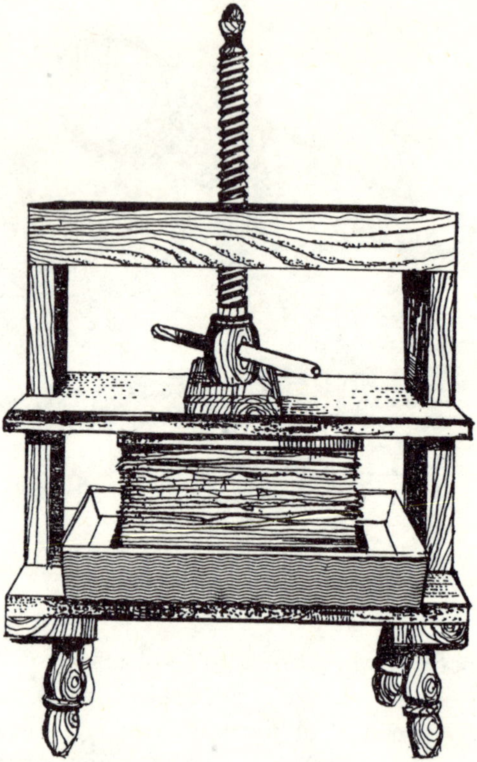

iron office letter-copying presses often can be had quite cheaply. Best of all for this work is a bookbinder's nipping or small standing press. This should be stood inside a metal trough and the whole bolted securely to a solid bench. If the trough is fitted with a water outlet pipe then you have the ideal piece of equipment needing no tedious emptying and causing no mess. The pressing of paper affects its quality and

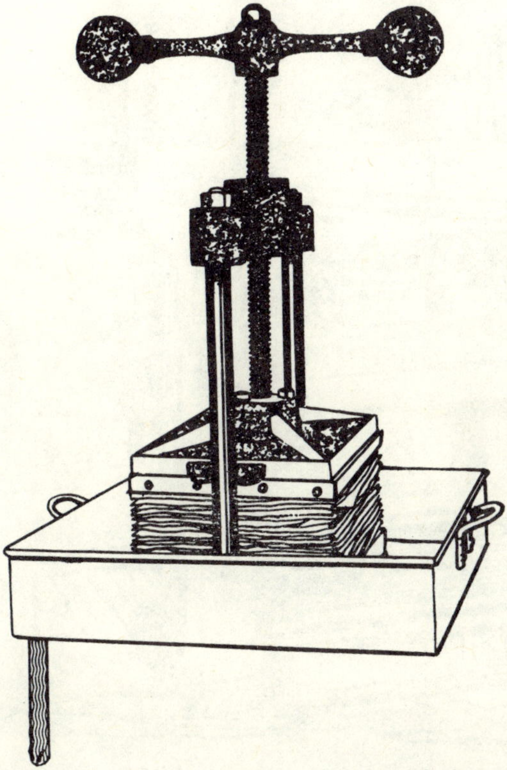

much will be learned by experiment and through experience. The size of press, to some extent, limits the size of paper which may be made, and future possible developments should be borne in mind when making a purchase. It is possible, of course, sometimes to press paper larger than the press itself if the pile is inserted the narrow way round. Then you may need to screw down at first one end of the pile and then the other.

Old press in use at Richard de Bas Mill, Auvergne, France

The wooden pressing boards and blankets both should be 4 inches larger each way than the largest paper you intend to make. The more water resisting, the longer the wooden boards will last, and I do strongly recommend teak.

At first, however, use whatever you can get. The really important thing is to get going somehow as quickly as possible and to make improvements later.

Pressing Equipment

Paper makers' felts may once have been made from tightly compressed unwoven wool or hair. Nowadays, however, though the term 'felt' is still retained, a specially woven woollen blanket material is used. This has strong, combed and well-twisted wool for the warp to give long and hard wear, but the weft is of carded wool more loosely spun to give good drainage. The weaving, too, is fairly loose, for a tight weave also retards the removal of the water. New felts have a rather high-riding nap which makes couching more difficult, and so

those that have been in use for a while are much to be preferred. At first I used all sorts of household materials such as old well-worn bed blankets, stout sheets, table-cloths and curtains, and these gave quite interesting textures to my paper but, unfortunately, after some time they frayed badly at the edges and wore into holes. Eventually J. Barcham Green kindly let me have a good supply of small felts cut from worn-out larger ones from Hayle Mill and these have served me ever since. Paper-making machines use long endless felts and in those making newsprint they are of sufficiently loose weave to be suitable for hand work. Machine felts are discarded fairly frequently and, where available, should be cheap enough. There are, of course, suppliers of new felts for hand-made work, but these are quite expensive and, until well worn with use, the high nap can be very troublesome to a beginner. The nap of a felt is longer on one side than on the other, and it is an advantage when paper making to lay the long nap side downwards on the pile so that the sheet is couched upon the short nap side. There is considerable scope for experiment in the matter of materials for felts, and there may be some yet untried which would prove very satisfactory and impart new and interesting textures to the paper. The edges of all felts should be overlock sewn to prevent fraying.

CHAPTER VII

Forming the Sheets and Couching

Dipping laid mould and deckle

The professional paper maker rarely makes sheets less in size than Royal, or 20 × 25 inches. For smaller sizes either his mould or his deckle is divided up so that he makes two or more pieces at each lift. Special difficulties arise when making large sheets and the knack can be acquired only after long practice and by much experience. Considerable physical strength is needed, too. It is very different when making separate sheets about the size 12 × 8 inches. In an hour you should be making excellent paper. I expect that the very obvious difficulties seen in a paper-mill have previously discouraged any from practising this craft for pleasure. Private printers and private bookbinders have long contributed richly to their crafts and I believe that private paper makers, by their very limitations, could well do so, too. So defer a visit to a hand-made mill until you have become a

Dipping a cloth covered mould without deckle

A sheet of perforated zinc nailed to a wooden frame makes an excellent mould for thin sheets. They may be couched in the usual way. For fine pulp the zinc could be covered with nylon net

maker of small sheets yourself. Then you will be confident and understand better what you see.

Fill your vat with warm water to 2 or 3 inches from the brim and then add two jugfuls of pulp. Stir this briskly with a whisk, wooden stick or bamboo cane. Run your fingers through the mixture and, if still very fluid, add more pulp. When you judge that things are right it is time to make a test sheet. To the left of the vat should be a large photographer's developing dish, or something similar, to catch the water, and in it place a wooden pressing board. At the back of the dish stand a pile of well-wetted felts. Lay two of the wet felts on

Lifting and draining

the pressing board and smooth flat with the hands. Place the deckle over the mould and pick up both together with a hand on each side, fingers underneath and thumbs on top. Grip firmly just above the middle and hold vertically over the far side of the vat, with deckle and wire side of the mould nearest towards you. With a slow, continuous steady movement, dip the mould into the pulp and, when about 3 inches are immersed gradually carry it forward until the whole mould lies horizontal and an inch or two below the surface. Without

Couching

pause in this continuous slow movement you now bring the mould to the surface and give a slight sideways shaking movement from left to right and then forwards from back to front in order to felt the fibres. It is this shaking movement in both directions which contributes so much to the special strength of hand-made paper. Though this movement is very slight and easily made with small sheets, it is much more definite and very much more difficult to do properly when making large ones. Still maintaining a tight grip on deckle and mould, tilt to one corner and drain until the water has nearly ceased to drip. If the film of pulp on the mould, now known as the waterleaf, is quite obviously too thin, then return it to the vat by turning the mould over and laying the wet pulp in contact with the water. A sharp lift and the pulp is left to be stirred back into the vat. Stir in some more pulp and make further tests. When you have a layer which appears to be at least a sixteenth of an inch thick, then it should be couched. Thinner sheets are more difficult to make and to handle and should

Mould lifted

Sheet couched to felt

not be attempted until later. Remove the deckle and stand it to drain at the side of the vat. Turn the mould upside down ready for couching, the sheet will not fall off. Now place the near edge of the mould on to the felt and with a firm rolling movement press the whole surface down and so transfer the sheet to the felt. This continuous rolling movement, similar to that used with the old half-round type of desk blotter, is soon

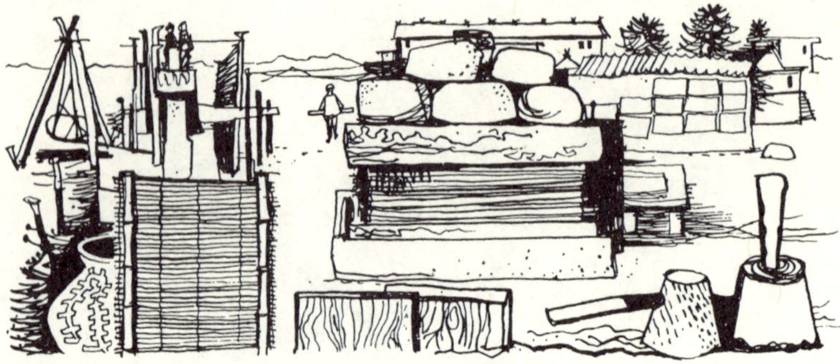

perfected. This, and the actions of dipping and lifting the mould, should be practised until they are easy and rhythmic. Keep the hand grips firm but the arms quite loose. Now lay a felt on your newly made sheet, taking care to keep all edges level. Stir the pulp well, for the fibres tend to settle, and then make a second sheet as you made the first. Remember that each time you lift a sheet you thin the mixture. Roughly you remove a small cupful of pulp each time. It is sufficient if you add the equivalent of six cupfuls after every six sheets are made. When everything is going well it is usual to make a pile or 'post' of alternate felts and sheets equal to the capacity of

your press. Then, with an extra felt and wooden board at top and bottom, the pile is placed in the press and slowly squeezed tighter and tighter until it reaches the maximum of the press. Rub your hands round the edges of the felts to press out as much as possible of the water. It is not necessary to wait until you have a full post before pressing, a few sheets or even one only may be treated similarly.

While you are waiting, thoroughly wash mould and deckle and stand up to drain. Also mop up any slopped water.

After a minute or so loosen the press and remove the pile. Take off the top board and the first two felts and there smooth and flat lies the first sheet of your paper.

CHAPTER VIII

Separating, Wet Pressing and Drying

Wet freshly made sheets are very tender and need to be handled with delicate touch. My description of the trade method of separating, which follows, should be read but not actually tried out at this stage.

The layer or layman, as the worker is called, picks up the sheet by the corners of the near long edge, skims it off the felt and then lays it smoothly on a board. The next sheet is laid similarly, with all edges level, on top of the first and the process repeated until the pile is complete. The paper is now pressed again for a period to remove more of the water and to consolidate the fibres. This pressing also smooths the paper by removing most of the rough texture imparted by the felts. Further smoothing is obtained by repeatedly dividing up and rearranging the pile and giving further pressings.

There are several reasons why I suggest that this technique is best not used for the time being. In the first place it was developed for large quantity production of uniform smooth sheets to supply the demand of printers and stationers. They required standard materials which offered no impediment to writing or printing. We are not governed by these requirements and therefore, at any rate at first, we can keep the

A very delicate touch is needed for lifting wet sheets from felts

lovely texture natural to the process. Also, it is very difficult to lay a pile of wet sheets quite evenly and without creases. The following method will be found to facilitate separating, to speed up drying and to preserve the texture.

Separating, Wet Pressing and Drying

Herm Island. The Agave or Century plant, so called because it can take up to sixty years to flower. The tough fibres are excellent for paper making

Cut up a quantity of strong stout white paper for lay sheets ½ inch larger each way than the size of the paper which you are making. A good thick cartridge paper will serve, but it wears out fairly quickly. I use Linson vellum because, though more expensive, it outlasts all other materials. My own sheets have been in almost daily use for over four years and, though now a little discoloured, are still doing their job as well as ever.

Let us return to the pressed post with top felts removed and first sheet of paper exposed. Obtain a fairly thick sheet of glass about 2 inches larger each way than your paper. Place a lay sheet over the first wet sheet and peel both off the pile to-

Lifting with Linson carrier sheets

Hardboard racks or wire mesh trays facilitate drying and are compact. Collect the sheets before they are bone dry and place under a weight to flatten. Tissues and other thin sheets and also those containing such additions as fern leaves and grasses are best left on the blankets to dry

gether with, and by means of, the first felt immediately below. Turn the whole upside down and lay flat on the glass. Peel off the felt and lay this to one side. Repeat this operation until you have all the sheets in the post piled alternately with lay sheets, and all the felts removed. It is now quite simple to spread out each sheet separately to dry together with its lay sheet below it. This method obviates the handling of the wet sheet itself and prevents any risk of marking, tearing or stretching it. Later you may need to make thin paper and then this method needs a slight modification. While paper of fair substance dries perfectly laid out in single sheets, thin paper shrinks and often develops bad cockles. For this reason thin sheets are grouped for drying in 'spurs' of four or five sheets at a time. Each one may be picked up from its felt by a lay sheet, but only the first in each group of five retains the lay sheet. The others are carefully placed upside down on the previous sheet and the lay sheet, like the felt, is peeled away. The paper may be laid out to dry on any smooth clean sur-

In the East sheets of paper are spread on to mud walls and left to dry in the sun. We can use an electric convector heater with fan to speed up the drying

Sinkiang peasant bringing a load of mulberry and bamboo for paper making. In Tibet floors of dwellings are often made from paper

face. My own drying racks have sliding hardboard shelves which allow for handling in quite large numbers and in comparatively small space. Trade drying of their larger sheets is carried out in well-ventilated lofts and at one time the spurs were hung on cowhair ropes because these left no stain or mark.

Collect the paper from the lay sheets when all dampness has gone, but before they are bone dry, and place overnight between boards and under a weight to flatten. Finally, they should be pressed for several hours between stout pieces of white card.

Provided that the work has been done carefully these first sheets should give you a pleasant sense of achievement. No doubt they will have a coarse homely texture, pleasant to see

and to handle and, according to what raw materials you used, remind you of oatmeal biscuits or Harris tweeds. There may be too many coarse lumps and odd pieces for your liking. This may be remedied easily. The paper is too thick or too thin. The pulp can be adjusted accordingly. Writing ink spreads on your paper. That is cured by sizing. The aim was to make some paper without delay, full of faults no doubt, but enabling you to learn at first hand, and in a practical fashion, the basic principles which underlie all paper making.

At greater leisure now we can consider some additional processes, some variations and perhaps a few refinements.

It is quite pointless for us to try to imitate any of the present-day rather dull, flawless, machine-made papers. Excellent white rag hand-made printings and writings, too, of better quality than we can ever hope to produce may be bought from the existing trade mills. We, however, have completely free hands. Let us make capital of our primitive methods and

equipment, bringing imagination and new ideas to this ancient craft, and perhaps reviving and helping to preserve something of its early beauty.

Some beginner paper makers recycle computer tape and other types of paper in a liquidizer such as that shown on page 93. It must be remembered that such paper has far less strength than the original sheets from which it is made. It may have some small use in facilitating the demonstration of sheet forming or as an economy measure but, otherwise, is not to be recommended.

CHAPTER IX

Sizing

Paper made, as already described, from garden and field plants, is quite suitable for printing without any further treatment. That made from cotton rags, however, is more absorbent and requires a light sizing. It is generally considered better to make paper unsized at first, allow it to mature and then to size it as a later operation. This separate tub sizing as it is called, undoubtedly is needed for writing paper. For printing paper made from rags, however, provided that the vat can be heated, the size may be added to the pulp to save the separate operation.

Into a bucket place a measured quantity of good skin glue, preferably in powder form, cover with cold water and leave to soak overnight. When jellied, add boiling water and stir well. Take a measure of chrome alum, weighed to about one-tenth of the dry weight of the size, dissolve in water and add to the solution. The alum improves the action of the size and prevents the growth of fungus or mould. Where an electric heater plate has been fitted to the bottom of the vat, a quantity of the size may be added to the pulp. At first you will need to make and dry a test sheet to discover the correct quantity to add. Very soon you will be able to judge this without test-

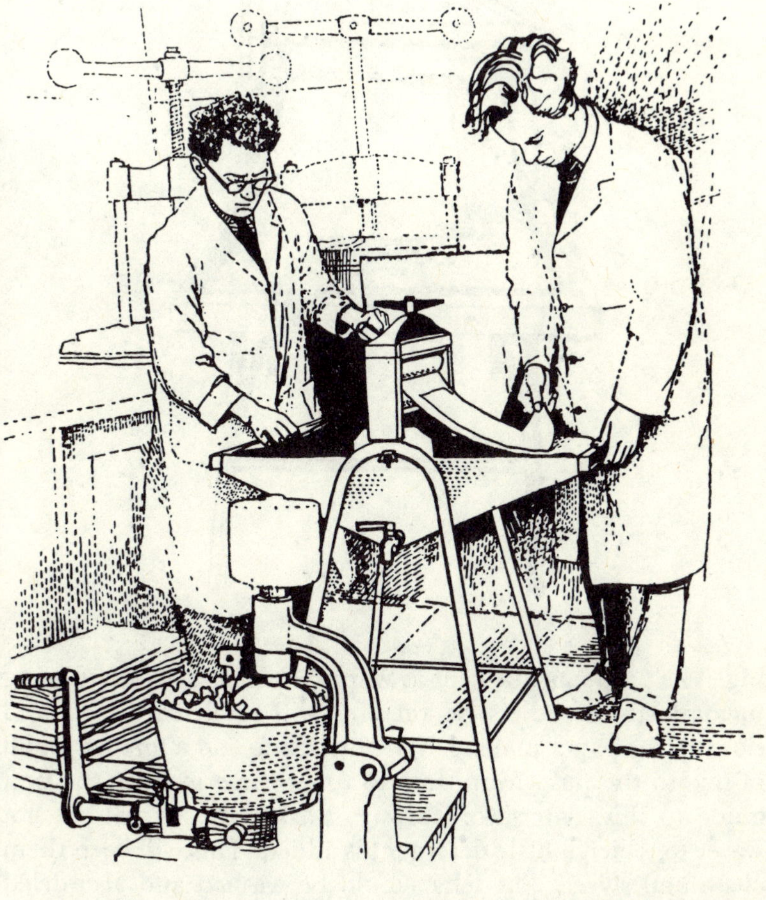

A sizing machine adapted from an old Acme wringer by Roland Harvey. An endless felt revolves in a bath of size. A sheet of paper is laid on at one side, absorbs the size, passes through the rollers and is removed on the other side

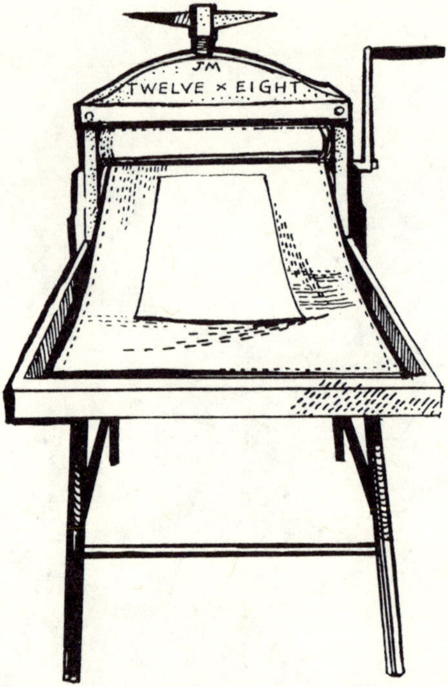

ing. The pulp must be kept as warm as possible, without being uncomfortable, otherwise the size will retard draining and slow up the paper making. When working add a further jugful of size to the pulp from time to time. Experience is the best guide to this. After work always wash the used felts in hot water to which a little detergent is added. This will keep them clean and sweet. The felts should be washed and also dried when there are long intervals between periods of working or they will soon have a most unpleasant smell.

Never tub size sheets until they have been made for at least two or three weeks, otherwise they are likely to disintegrate in the bath. Pour warm size into a developing dish or copper tank, and in it place a sheet of glass. Immerse a sheet of the unsized paper and with the hands gently push it down to the glass. Immerse a second and carry it down so that it lies exactly on top of the first. Continue until you have a pile of

In the East mucilage is added to mulberry and hibiscus pulp. This acts as a lubricant for the exceptionally long fibres in much of the paper

twenty-four sheets. Lift glass and paper from the bath, drain, lay a felt on the top sheet and then a wooden board. Turn the pile over, remove the glass and, at this side also, lay on a felt and wooden board. Give a light quick nip in the press, separate carefully and place on your lay sheets to dry. When dry, give a final pressing. Too strong a size will prevent parting of the sheets. Two separate sizings in a thin solution are far better than one sizing in a thick solution.

There is still much to discover about paper sizing. When making their lovely mulberry papers, the Japanese add a special vegetable mucilage to the pulp. This not only serves for size but prevents entanglement of the exceptionally long fibres. It is extremely difficult to make such papers by our Western methods.

Above Hokeo, Tibet. Bamboo being split and stacked preparatory to making paper

CHAPTER X

Decorative and Straining Devices

Many unusual papers may be made by mixing different types of pulp together in the vat. Also blends of the same fibre with varied degrees of beating often result

in most unexpected and exciting effects. Trade coloured papers have the dye added to the pulp while still in the beater. Many experiments over a long period have convinced me that the wise small paper maker will leave dyes quite alone. Unwanted colour gets everywhere, into felts, materials and equipment. It is most difficult to remove even with most careful cleaning. By far the best plan is to beat up already dyed cotton rags and to keep the different colours in separate jars. Pulp in the vat may then be tinted by the addition of a small quantity from one of the jars. For papers of deeper colour you will need to use the coloured rag pulp alone. Quite beautiful effects may be obtained by placing a small jugful of each colour into the vat and giving only a very light stir. The resulting paper has a charming effect reminiscent of Scottish tweeds.

Coloured materials, plain or patterned, may be cut up with scissors into tiny fragments and used for tinting rag pulp. The intermingled tiny hair-like flecks give very lovely colour effects usually known in the trade as 'silurian'.

Skeleton leaves, flower petals, grasses and coloured threads may be sandwiched between two layers of waterleaf. Some Japanese papers even incorporate butterflies

If a sheet of paper is couched to a felt and then another couched immediately on top, the two sheets will bond and form one thicker sheet. When two vats are used for this twin-couching then paper may be made of different colour on each side. If the second couching deposits only the merest wisps of fibre, these will form an almost transparent layer. This layer may be used to seal-in all sorts of decorative devices applied to the upper surface of the first full thickness sheet. Fill a vat with water and stir into it a tablespoonful of well-beaten cotton pulp. Add more pulp only until you are able to lift the very lightest gossamer layer upon your mould. There is no need to use the deckle for this very thin layer. Couch a sheet of normal thickness from your first vat in the usual way. Then, immediately, lay on the top of this some short lengths of thin coloured cotton threads, pressed flowers or skeleton leaves, all of which need first to be wetted. Couch the fixing sheet, cover with felts and boards and press in the usual way. You will be able to develop many variations of

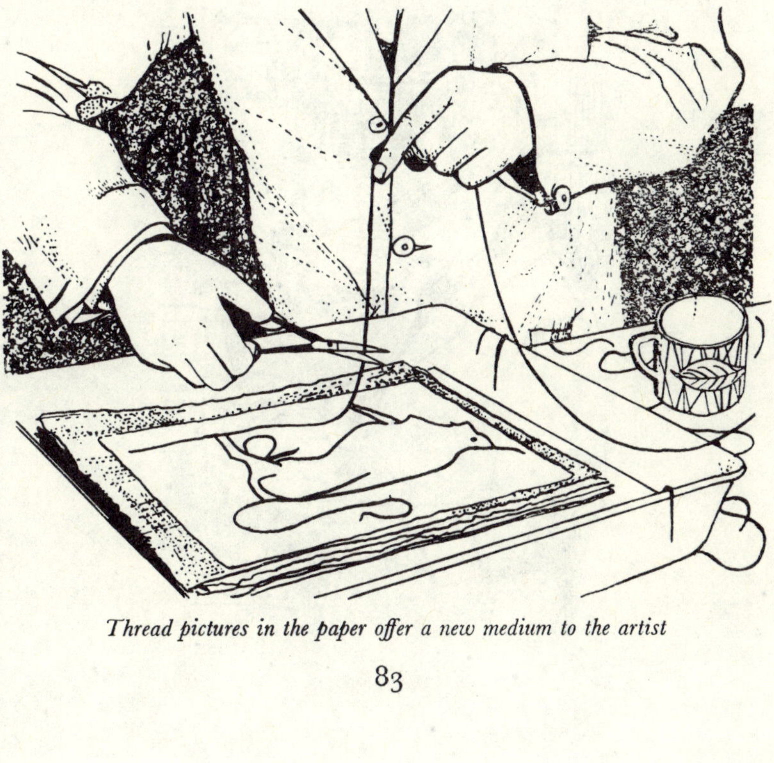

Thread pictures in the paper offer a new medium to the artist

this technique. After a little practice continuous coloured cotton threads can be laid in a controlled way to form patterns, pictures and even lettering. Laurel and magnolia leaves and probably many other varieties, too, may be skeletonized by boiling in a solution of caustic soda. When the fleshly portions are sufficiently softened and loose, lay the leaf on a sheet of glass, hold under a running tap and carefully brush until only the cellulose skeleton remains. When dry, such skeleton leaves may be dyed, if desired, by dipping into various coloured inks.

Lumps and coarse pieces in the pulp, often attractive in

The ball sieve is a valuable strainer

individual paper making, sometimes are a nuisance. In the trade the pulp passes through a special type of strainer before reaching the vat. An ordinary sieve, even a coarse one, is useless for this purpose as it clogs almost immediately. This prob-

Simple hot pressing

lem bothered me for a long time before I discovered an answer.

One day, in my larder, I found hanging by its handle a ball shaped wire sieve salad-shaker hinged to open in half as shown in the drawing. This I requisitioned and it proved to be the solution. I placed two or three handfuls of pulp into the two halves, snapped it shut and then plunged it up and down vigorously in a bucket of water. The upward pulls caused the water to suck the pulp through the wire, leaving only the hard

A variety of steam, charcoal and flat irons

In India the paper is smoothed or burnished with large stones which gives a very attractive glazed appearance and facilitates writing

I sometimes use a Kodak print glazing machine for imparting smooth finish

lumps behind. I repeated this operation until I had sufficient strained pulp for my paper making. Unfortunately, this type of salad shaker is no longer on the market. It has been replaced by one made of plastic and with openings much too large for our purpose. A handy tinsmith should be able to make some sort of strainer box with handle without great trouble. Such a tool is invaluable and is best made of stainless steel. Otherwise it is advisable to have it plated to prevent rust, that arch enemy of the paper maker.

One last hint and I have done. Hard sized sheets made for

writing or calligraphy should finally be smoothed on both sides with a hot laundry iron. Then you will have true 'hot pressed' paper. Parcel up your sheets, let your stocks grow and mature and then, one day, when you wonder what to do with them, I suggest that you read John Ryder's *Printing for Pleasure*.

I have enjoyed nothing in my life more than these adventures into paper making. It has been pleasant, too, to write this book. In conclusion, I hope that now others also will try their hands at this delightful and satisfying craft.

Rigby Graham

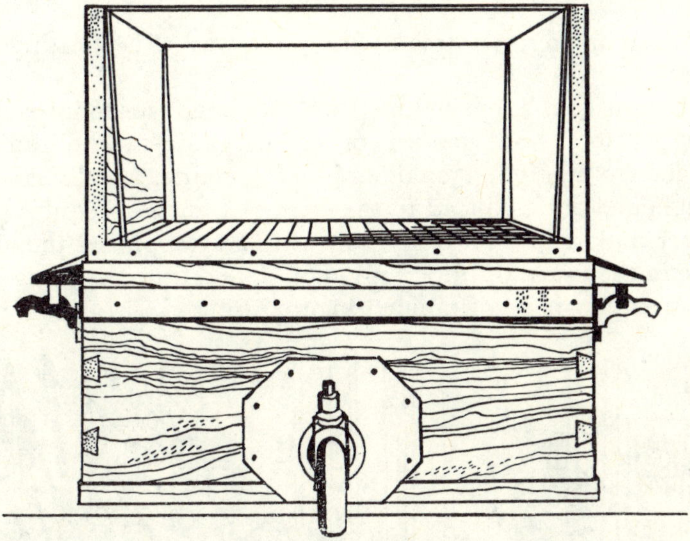

CHAPTER XI

A Note on my Nylon Paper

I have received so many inquiries about my recent experiments with synthetic fibres that I feel the following short account of this work may be of interest.

In 1955 I started to experiment with nylon. I could get none in staple form at that time, and so had to unwind filament yarn and chop down to $\frac{3}{8}$-inch lengths with a guillotine knife. This multi-filament staple yarn would not disperse in water so I boiled it in a weak solution of Tepol.

Into my beater went the nylon, but it only broke up the filament and nearly broke my small beater. Long experiments followed in my attempts to find some way of separating the fibres.

Eventually I subjected my staple to a high-speed impact in a laboratory homogenizer or disintegrator at the rate of 12,000 r.p.m. giving 50,000 impacts a minute. My fibres were roughened and reduced to the degree of fineness required for paper making. I mixed my pulp with water, placed the mixture in the vat and dipped my mould, and fished nothing but an unsightly mass of tangled knotted lumps.

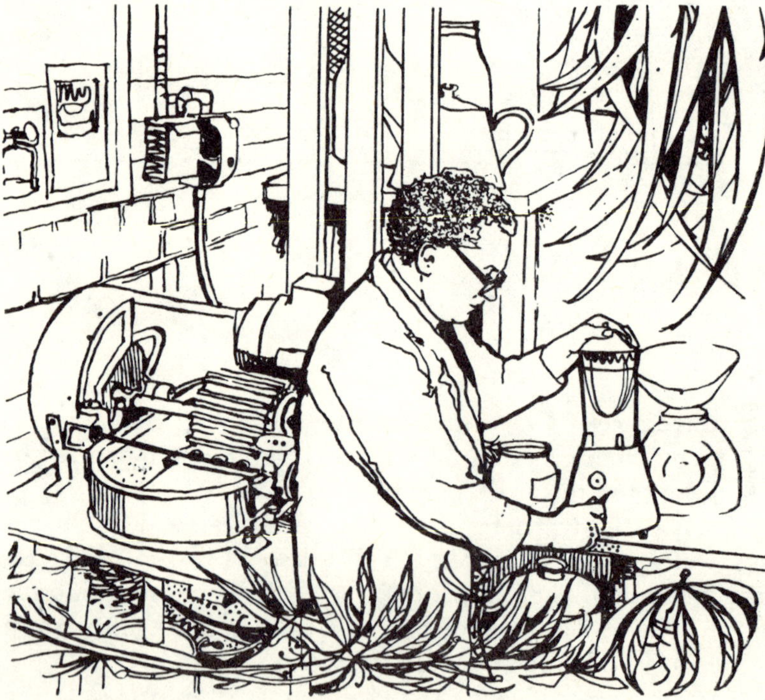

High speed disintegration

A Note on my Nylon Paper

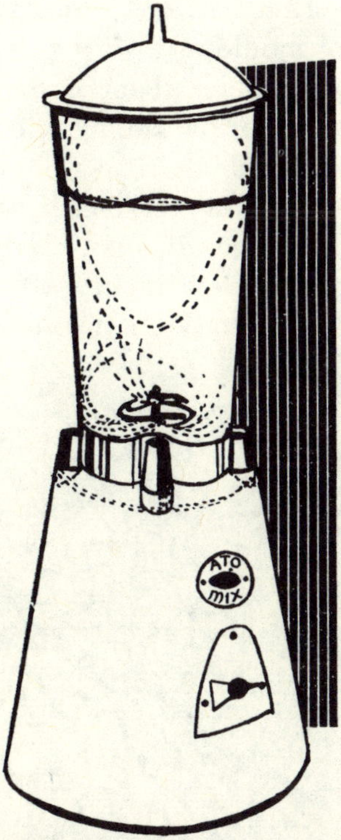

More experiments with dispersing agents followed until I found the right one in Cellofas B. But now I had a viscous fluid in my vat and dipping the mould was useless. So I built a cabinet, like a small tank, to take my mould and with an emptying tap below it. In went the mixture, the tap was turned and I had a lovely even glistening wet sheet of nylon waterleaf. When dry the sheet looked like a layer of matted wool. A final process of pressing under heat and I had made my first sheet of nylon paper.

Tests proved the sheet had enormous strength and resistance to the action of mould and strong chemicals. Here was the ideal imperishable material for documents and records and for paper currency, for packaging of corrosive substances and who knows what else.

Later I made paper from terylene also. Existing paper-making machines will not handle man-made fibres in this way successfully, however. New machinery will have to be devised before this type of paper can be made on a commercial scale.

Glossary

BEATING. The operations involved in the reduction of raw materials to the fineness required for paper making.

COUCHING. Transferring the newly made sheet by pressure to a wet felt.

DECKLE. Loose wooden frame which confines the pulp upon the mould. The rough edges on hand-made paper caused by this loose fitting are known as deckle edges.

DHENKI. Special type of beater used in India.

FELT. The blanket on which the sheet is couched.

HOLLANDER. Another name for the 'beater' which was a Dutch invention.

HOT-PRESSED. Paper with surface smoothed by hot plates, rollers or by ironing.

HOG. The mechanical device fitted at the bottom of a professional vat for preventing the fibres from settling to the bottom.

LAID. Moulds formed by fine cross wires divided at intervals by lines of link stitching. Laid paper is watermarked by these lines as distinct from smooth wove paper.

MOULD. The rectangular sieve or strainer on which the sheet is formed.

POST. The pile of newly made sheets between felts ready for pressing.

RETTING. Partial disintegration by rotting.

SILURIAN. Paper having numerous minute flecks of fibres of other colours. I wish that I knew why it is so called.

TUB SIZED. When paper which has been dried and seasoned is subsequently immersed in warm size as a later operation. Usually used for writing papers.

VAT. Any container holding the pulp for paper making. The workman who stands at the vat to form the sheets is known as the vatman.

WATERLEAF. The newly made wet unsized sheet.

WATERMARKS. Devices stitched to the surface of the mould and which are reproduced in partial transparency in the paper made on it.

WOVE. Mould carrying fine woven wire and which forms smooth paper without the watermarked lines found in laid paper.

Index